Genetics and You

Genetics and You

John F. Jackson, MD

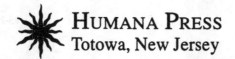

HUMANA PRESS
Totowa, New Jersey

An earlier version of *Genetics and You* was issued directly by the author. Its success in science education courses demonstrated the potential value of the book to general lay audiences, and strongly encouraged the author to seek out a trade publisher for this present updated and enhanced version.

© 1996 Humana Press Inc.
999 Riverview Drive, Suite 208
Totowa, New Jersey 07512

This publication is printed on acid-free paper. ∞
ANSI Z39.48-1984 (American National Standards Institute) Permanence of Paper for Printed Library Materials.

Printed in the United States of America. 9 8 7 6 5 4 3 2

Library of Congress Cataloging-in-Publication Data
Jackson, J. F. (John F.), 1935–
    Genetics and you / John F. Jackson.
        p. cm.
    Includes index.
    Summary: Provides the nonprofessional with an awareness of why the risk for genetic disease may be so high and does so by explaining genetic principles in general.
    ISBN 0-89603-329-5 (hardcover : alk. paper). —
ISBN 0-89603-330-9 (paperback : alk. paper)
    1. Medical genetics—Popular works. 2. Genetics—Popular works.
[1. Medical genetics. 2. Genetics.] I. Title.
RB155.J33 1996
616'.042—dc20   95-20569
                                                          CIP
                                                          AC

Foreword

Familiarity with the basics of genetics is something that everyone desires and requires—certainly anyone who is having children, anyone who has a potentially genetic problem in the family, and anyone who is interested in the exciting initiative to map all 60,000 to 70,000 human genes, the Human Genome Project. And that is just about everyone. As Dr. Jackson, the author of this book, points out, no one can help but be intrigued by questions of why close relatives are so much alike—or so much different.

In the last 30 years during which Dr. Jackson has been working in the field, human genetics has become heavily medicalized. The specialty of medical genetics has been established. At the same time, a democratization of human genetics has taken place. The results of genetic research related to the human have been of ever increasing general interest. In the case of genetic diseases and medical genetics, consumerization has occurred as well. For example, essentially every genetic disorder and birth defect has its own support group. These groups serve a highly useful role in accumulating and disseminating information on specific disorders and on genetics in general. The Alliance of Genetic Support Groups coordinates about 250 such organizations.

Sources of information on the basics of human genetics and medical genetics for the general public are not numerous. Authoring such sources requires knowledge, writing skills, and experience, both in the practical aspects and in communication. John Jackson fulfills all three prerequisites. As a result, this small book provides the basic genetic information that everyone desires and requires.

Victor A. McKusick, MD
Center for Medical Genetics
Johns Hopkins Hospital
Baltimore, MD

Preface

Children resemble their parents. Relatives look for and comment upon likenesses with statements like: "He's got that same turned-up nose like his grandfather," or "Her mother had blonde hair when she was a child, too." People take for granted the inheritance of physical features and seek out family likeness—usually with pleasure.

On the other hand, questions about less desirable traits such as: "Do you think junior is going to be bald like his father?" also arise. Some might think baldness is a trivial matter. Because baldness is so common, though, there was recently a change in the price of a drug company stock based on expected sales of a newly approved medicine for the treatment of baldness. Most of the inherited disorders that worry parents and may require the services of a medical geneticist are individually not nearly so common as baldness. But genetic disorders jointly comprise a large burden of human suffering; and many, of course, are very serious disorders.

There are more than 4000 inherited traits and the chromosome location is known for over 1000. Up to 4% of newborn infants suffer from a major congenital defect. Congenital simply means present at birth without regard to the cause. At least one-fourth are due to the combined effects of multiple genes plus one or more environmental factors (multifactorial), but almost half of all birth defects do not have a known cause. About 1 in 166 newborns has a chromosomal abnormality such as Down syndrome. Some 2–3% have disorders caused by a single abnormal gene. Probably 20% of pediatric hospital patients have a problem that is partly genetic. Congenital abnormalities are the third most common cause of death for ages 1 through 14, following only accidents and cancer. More than 20% of infant deaths are caused by birth defects. Thus the burden of human suffering from genetic disorders is huge.

People frequently remark that "Cancer runs in our family," or "The men in our family all die of heart attacks." Since heart disease and cancer are the two leading causes of death in the United States, the real question is whether or not there is a risk in any given family that is greater than that in general. In some types of cancer, for example, the risk for children of affected persons also to develop cancer is one in two.

Details of human genetics grow so rapidly that even the professional medical geneticist has trouble keeping up with the huge amount of information. In addition, new technology constantly provides fresh ways to understand and manage genetic problems. The basic principles of genetics are few, however, so that a small number of rules fits a host of individual problems, It is the aim of this little book to provide the nonprofessional with an awareness of why the risk for genetic disease may be so high—by explaining genetic principles in general. The result should allow questions about inherited disease to be properly directed, answered, and understood. In short, the goal is to provide basic information about genetic disorders in a useful form for those who have no special training in genetics.

It is not the intent of this book to provide "do-it-yourself" genetic counseling, but to assist those who have questions about genetics to find help. Expectant mothers can read more about why their obstetricians are offering them prenatal testing. Couples who get counseling for genetic disorders can refresh their memories when the question arises: "Why was it that the doctor said our risk to have another affected baby is one in four?" Hopefully, others may simply be curious to learn about current use of genetic principles and modern technology in human genetic disorders. I happen to think that learning about why we are the way we are is fun; but after 30 years in medical genetics, that attitude should be expected from a person like me.

The reader can look up big words in the little glossary at the back of the book.

ACKNOWLEDGMENTS

Thanks to Georg Bock, Robert Brent, Wayne Finley, and Stephanie Smith for helpful suggestions. William Buhner produced the computer graphics and John Gibson provided Fig. 13.

John F. Jackson, MD

Contents

CHAPTER 1

Principles of Genetics

CHROMOSOMES, GENES, AND DNA

The chromosomes are the packages of genes that allow exact passage of genetic information from one generation to the next—from cell to cell as tissues grow, and from parents to child as families grow. When the cells are doing their thing—brain cells thinking and liver cells storing glucose, for example—the chromosomes are so uncoiled inside the cell nucleus that they are not visible. It is only during the process of cell division when the chromosomes shorten by coiling up that they can be seen as separate bodies.

Figure 1 is a photograph of the chromosomes from a single cell of a normal male. Live cells can grow in culture medium in a laboratory flask. After stopping cell division (called mitosis) with a chemical and swelling the cells with a dilute solution that causes them to take up fluid, the dividing cells are spread onto microscope slides. We can see the untangled chromosomes using an ordinary light microscope, since the largest chromosome is about as long as the width of a red blood cell. The chromosomes have doubled in preparation for cell division. Had cell division continued, each chromosome in the photograph would have split

A

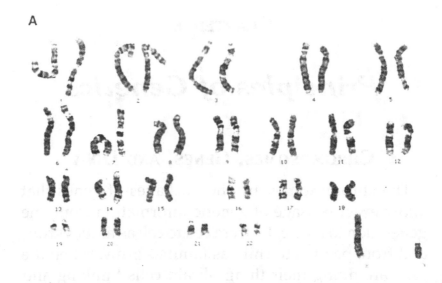

Fig. 1. **(A)** Chromosomes of a male cell arranged to form a karyotype, representing the chromosome set for that person from whom the white blood cell was taken.

lengthwise down the center. One half (a chromatid) would go to each of two daughter cells, giving a complete set of genes to each new cell (*see* Fig. 2).

Modern staining methods show a series of dark and light bands that identify each chromosome. Figure 1B is the picture of a cell as it was seen in the microscope with the chromosomes scattered at random. Individual chromosomes cut from enlarged photos and arranged in order from largest to smallest produce what is known as a karyotype, shown in Fig. 1A. The karyotype shows the chromosome set of the person from whom the cell was obtained. Television cameras attached to microscopes allow computer sorting of stored images and laser printing of the results, avoiding the time-

B

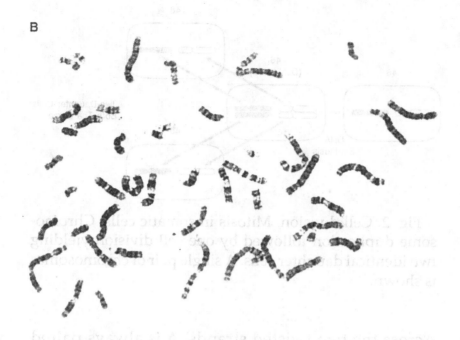

Fig. 1. **(B)** Photograph of the chromosomes in a single cell from a normal man. This is how the cell looked in the microscope before the chromosomes were arranged in order from largest to smallest to produce the karyotype in Fig. 1A.

consuming steps of photography. The computer method was actually used to obtain the photograph in Fig. 1A from the cell shown in Fig. 1B.

The individual genes are lined up in order on each chromosome like beads on a string. They are too small to be seen in the microscope, and even the smallest band seen in the picture consists of many genes. The genes are composed of deoxyribonucleic acid, usually abbreviated DNA. The DNA of a single chromosome is a continuous double strand made of long strings of the building blocks adenine (A), thymine (T), cytosine (C), and guanine (G).

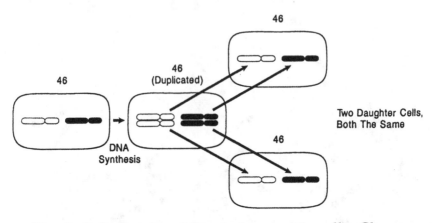

Fig. 2. Cell division. Mitosis in somatic cells. Chromosome duplication followed by one cell division yielding two identical daughter cells. A single pair of chromosomes is shown.

Across the two twisted strands, A is always paired with T and C with G, as in the following example

TATACGTTCCGATAC

| | | | | | | | | | | | | | |

ATATGCAAGGCTATG

When DNA duplicates prior to cell division, the strands pull apart like a zipper. New nucleic acids paired C to G and A to T with the two old single strands result in two new double strands with the same order as before (Fig. 3).

The genetic code uses a sequence of three-letter combinations of A, T, C, and G along one strand to specify the building blocks of protein, the amino acids. The code is transcribed to ribonucleic acid (RNA) acting as a messenger to tell the cell machinery how to translate the message into protein.

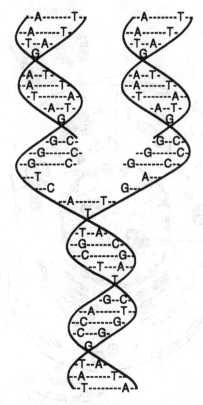

Fig. 3. Original strand of DNA below is separating and two new identical strands are being formed above by adding the correct bases: G with C, or A with T.

Some genes code for what are termed structural proteins, such as the hemoglobin molecule in red blood cells that carries oxygen from the lungs to the tissues. Other genes specify the protein structure for enzymes that carry out the additional chemical reactions to be performed by specific cells under given conditions—in short, whatever functions are provided for by inherited means.

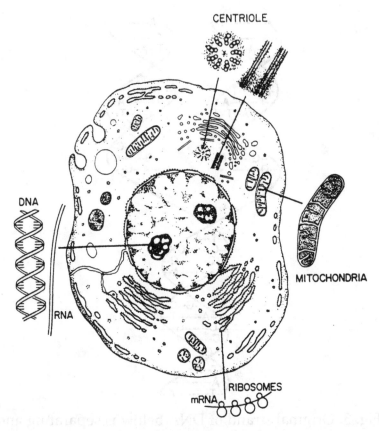

Fig. 4. Diagram of a typical cell showing a round central nucleus containing the chromosomes with DNA in an extended form, and RNA in structures called nucleoli. Other structures: centrioles, mitochondria, and ribosomes in the cell cytoplasm surrounding the nucleus are the cell machinery. The mitochondria provide the power for the cell, and were probably derived from inclusion of bacteria into cells early in evolution.

DNA also occurs in the mitochondria, small bodies outside the nucleus in the cell cytoplasm that serve to produce energy (Fig. 4). Mitochondria may have originated as bacteria incorporated into cells early in evolution. Mitochondria are plentiful in eggs, but are absent from mature sperm. All the mitochondria come from the mother's relatively large egg, and none from the father's sperm.

Studies using the DNA in these tiny mitochondria have been used to track maternal lineage searching for the earliest female—to look for the origin of "Eve"—in human evolution. Although mitochondrial DNA includes only 1% of total DNA, altered mitochondrial genes have been found to be the cause of several important human disorders involving nerve and muscle. When we talk about genes, though, we usually mean the 50,000 or more genes in the nuclear chromosomes of human cells.

SEX DETERMINATION—XS AND YS

The karyotype in Fig. 1 is that of a male, since the two sex chromosomes consist of an X and a Y. Females have two X chromosomes. The rest of the chromosome pairs 1–22, termed autosomes, are the same for both sexes. Genes on the smaller Y chromosome determine maleness, but the exact mechanism is still incompletely understood. Only a very small segment of the tips of the short arms of the X and Y chromosomes carries homologous genes (genes for the same function). X linkage (sex linkage) is the term used to describe genes on the X chromosome not paired by the Y chromosome. Any abnormal gene on the X chromosome of a male that is unpaired by his smaller Y chromosome will be expressed because it is the only gene available for that purpose. Since the female has two X chromosomes, an abnormal gene on one of her X chromosomes in the presence of a normal gene on the other X chromosome may or may not have an effect. That depends on whether or not the abnormal gene is dominant—requiring only one abnormal gene to be expressed—or recessive—requiring abnormal genes at that locus on both chromosomes for the trait to be displayed.

A female with an abnormal recessive gene on one X chromosome and a normal gene for that function on her other X is commonly referred to as a carrier. She herself is not affected, but half of her sons are expected to be affected, depending on which of her X chromosomes she gives each son. Such X-linked traits as color blindness and hemophilia are classically seen to be transmitted from affected grandfathers through their unaffected (but always carrier) daughters. The carrier daughters in turn have half their sons affected and half their daughters carriers (*see* Fig. 5, X-linked recessive gene transmission).

Females actually have only one X chromosome that functions in each cell. Early in the developing female embryo, one or the other X chromosome in each cell condenses to form what is called a sex chromatin body, or Barr body, and is then inactive in providing genetic information. The sex chromatin body can be seen in the nucleus of cells by using a microscope. Counting sex chromatin bodies provides the basis for sex screening tests using cells scraped from inside the mouth.

The selection of which chromosome is to be inactivated is random. Thus, in some cells, one X chromosome is active, and in other cells, the other X chromosome is the one expressing information. Short segments of both X

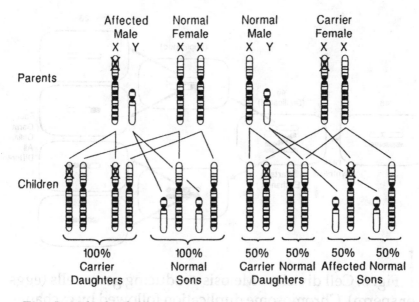

Fig. 5. X-linked recessive gene transmission. Abnormal gene(s) indicated by an X shown on the upper part of the X chromosome. Note that since a male who has the disorder gives his Y chromosome to produce males, none of his sons are affected because they cannot receive his X chromosome carrying the abnormal gene.

chromosomes remain active, but for the major part of the X chromosome, the gene products are equal in amounts to those in males. Thus, with respect to X chromosome gene action, all normal females are mosaics, that is, they have two different genetic types of cells. Sometimes a female carrier may show signs of X-linked genetic disease, depending on which and how many of her cells have the X chromosome carrying the abnormal gene as the active one. In individuals with abnormal numbers of sex chromosomes, only one X chromosome is active and all additional X chromosomes are inactive.

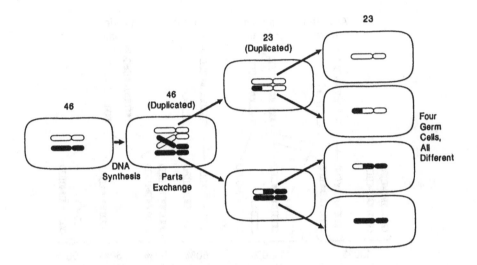

Fig. 6. Cell division. Meiosis producing germ cells (eggs or sperm). Chromosome duplication followed by exchange of parts and two cell divisions yields four germ cells, each different. Only one pair of chromosomes is shown.

THE GENETIC SHUFFLE

The production of germ cells (gametes)—the mother's ovum (egg) and the father's sperm—requires the exact division of the genetic material so that each parent gives half of his or her genes to each child. Since the genes are in the packages called chromosomes, this is produced by a special type of cell division termed meiosis (Fig. 6), in which only one of each pair of chromosomes is included in the germ cell. When the 23 chromosomes from the father join the 23 in the ovum from the mother, then the fertilized egg (a zygote) has the correct number of 46 chromosomes.

The random selection of one or the other of the chromosomes from each of the 23 pairs allows for great variety in the chromosomes of germ cells (2 to the 23rd power, or 8, 388, 608 possibilities for different combinations for each parent). In addition, the process of meiosis provides for even greater variety in sharing the inherited material. Prior to separation of the chromosomes into the germ cell, both chromosomes of each pair come together and exchange parts. This process of recombination, termed "crossing over," can produce an egg or sperm with any combination of available genes. It provides for literally infinite variety in inherited traits. Thus, the genetic makeup of every person is unique except for those individuals produced by splitting of an embryo to form identical twins from a single fertilized egg.

GENETIC LINKAGE

As a practical matter, the process of exchanging parts of the two chromosomes of a given pair during germ cell formation is limited. Therefore, genes that are close to each other on the same chromosome tend to remain together when they go to egg or sperm. This is termed genetic linkage.

Genetic linkage provides the basis for predicting the inheritance of genes that cannot themselves be detected, but can be expected to go along with a nearby linked gene that can be traced. It is the locus, or the spot where the gene resides on the chromosome that is linked to another locus, not the gene itself that is linked. The closer any two gene loci are on a chromosome, the greater the chance that the two genes at those two loci will go together to the next generation. Likewise, the farther apart that any two genes lie on a chromosome, the greater the likelihood that they will be separated and reshuffled by crossing over. There can be different genes for a given trait that occupy a given locus, so that linkage provides no permanent connection between any two specific genes. This is because the chance for separation of the genes at any two linked loci by recombination during germ cell formation occurs at each generation. By tracking the genes at linked loci through multiple family generations in many individuals, the probability of any two genes at those loci staying together can be calculated. Thus, genes at a linked "marker locus" can be used to predict the presence of a disease gene.

Detecting linked genes might be likened to looking for the contents of boxcars on a train. If we knew the cargo of successive cars on a train and we could recognize a given car by its color, then we could deduce the contents of another car by counting how far away it is from the known car. The order of cars would remain the same until switching occurred, when a car with different cargo might be added to the train. Gene switching also may happen when eggs and sperm form.

FLIPPING COINS

Except for X-linked genes in males, genes occur in pairs, so that most of the rules for inheritance derive from getting only one of each parent's two genes. The flip of a coin is a useful example, where heads is used to represent one gene and tails the other.

When a disease happens because only one gene of a pair is abnormal, that gene is termed dominant. A

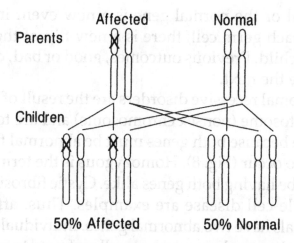

Fig. 7. Autosomal dominant gene transmission. Abnormal genes indicated by X. Each child of a person with a dominant disorder gets the abnormal gene (and the disorder) or else gets the normal gene, and therefore does not get the disorder and cannot pass the gene on.

couple of thousand or so human traits are autosomal dominant—autosomal referring to the gene being located on a chromosome that is not a sex chromosome (chromosomes 1–22 in Fig. 1). Huntington's disease and neurofibromatosis are examples of disorders inherited in an autosomal dominant fashion, which have recently attracted public attention. The "Elephant Man" has been thought by some to have had neurofibromatosis, but probably had a different disorder, Proteus syndrome. An affected parent will transmit autosomal dominant disease to each child with a probability of one in two—heads or tails, to use the coin analogy—because the affected parent must contribute one or the other gene, but not both (Fig. 7). Since the random selection of the chromosome with the

abnormal or the normal gene is a new event in producing each germ cell, there is a new flip of the coin
for each child. Previous outcomes, good or bad, do not
influence the next.

Autosomal recessive disorders are the result of genes
on an autosome (nonsex chromosome) and are termed
recessive because both genes must be abnormal for the
disease to occur (Fig. 8). Homozygous is the term used
to describe having both genes alike. Cystic fibrosis (CF)
and sickle-cell disease are examples. Thus, affected
individuals have two abnormal genes. Individuals with
only one abnormal gene are generally referred to as "carriers," since they carry the abnormal gene and can give
it to their children, but they are not affected themselves.
Heterozygous is the term for having different genes at
the same locus on a chromosome pair. The usual situation is for both parents of an affected person to be carriers. Since each parent must give one and only one of
his or her genes at that particular locus to each child,
then both parents, if unaffected, must be carriers to have
an affected child. The probability of subsequent children
being affected is one in four (Fig. 9).

In this case, it is like flipping two coins at once, one
for each parent. The possibilities are: heads and heads
(affected); heads and tails (carrier); tails and heads (carrier); and tails and tails (normal, noncarrier). Among
children of two carriers, then the expected ratios are 1
affected:2 carriers:1 normal. Again, each child repre-
sents a new flip of two coins, so that previous outcomes
do not alter the 1:2:1 risk. Figure 8 also illustrates the
transmission of recessive genes in other matings. An
affected person married to a carrier will have half

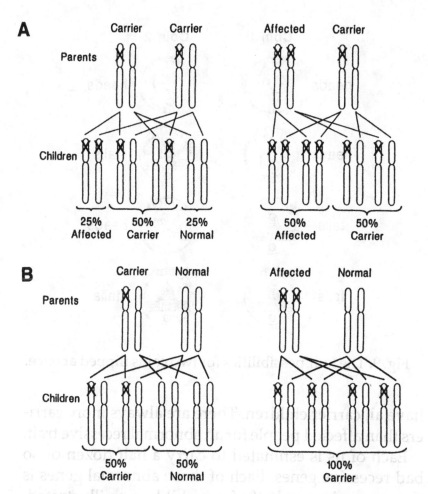

Fig. 8. (A,B) Autosomal recessive gene transmission. Abnormal gene(s) indicated by X on the upper part of the chromosome. One in four of the children of two carrier parents are affected by the disorder, as illustrated in A. As seen in B, carriers married to normals do not have affected children.

affected and half carrier children. A carrier married to a normal will have half carrier and half normal children, and an affected person mated to a normal will

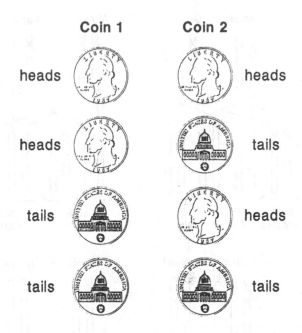

Fig. 9. The four possibilities for two coins flipped at once.

have all carrier children. There are always more carriers than affected people for an abnormal recessive trait.

Each of us is estimated to carry a half dozen or so bad recessive genes. Each of those abnormal genes is then passed on to half of our children as illustrated. Fortunately, most parents do not carry the same abnormal genes, so their children are unaffected. Sometimes both parents carry a different abnormal gene at a locus. In this case, a child can be affected by a disorder owing to receiving two different abnormal genes. This is termed compound heterozygosity. Hemoglobin SC disorder is an example.

Autosomal disorders are equally allotted to males and females since the genes involved are on the chromosomes common to both sexes. X-linked diseases have different frequencies in males and females, because males have only one X chromosome and females have two, as previously noted. X-linked recessive disorders generally affect only males, the exception being when a female has abnormal genes on both of her X chromosomes. X-linked recessive gene transmission is illustrated in Fig. 5.

There are a few X-linked dominant disorders. The transmission of dominant X-linked traits by mothers is one in two, just as in an autosomal disorders, since there are two genes on two chromosomes involved. Affected males have all affected daughters, however, because the daughters, in order to be daughters rather than sons, must receive their father's X chromosome carrying the abnormal gene. Likewise, in order for a child to be a boy, the father must transmit his Y chromosome, and therefore cannot give any X-linked gene—dominant or recessive—to his sons. That is not to say that a colorblind man cannot have a colorblind son. He can, but the son's gene for colorblindness must come from a carrier mother, not the father.

CUTTING THE CARDS

With our coin flipping examples, we have been illustrating the patterns of inheritance of single genes. Many common disorders are the result of combined effects of multiple genes and environmental factors acting togther (multifactorial). Congenital heart diseases (notice that I use the plural because there are many different types of congenital heart defects), spina bifida, and cleft lip are examples of common multifactorial disorders. For something like the development of the heart, which begins in the embryo as a single tube folding and then dividing into separate chambers and valves, it is logical that these complex events must require the action of multiple genes. In most cases, there is little if any information on which or how many genes are involved or on what external influences interact to produce the defect. Risks for diseases occurring again have been gathered in many cases simply by counting up how often the disorder affects other relatives.

Most times in multifactorial disorders, where there has been only one affected child, the risk for the next child to be affected is about 1–4%. The analogy in this instance then is that for a risk of 2%, it is like cutting a single deck of cards and turning up the ace of hearts (*see* Fig. 10). Since there are 52 cards in a deck, then the chance of picking any card is about 1 in 50 or 2%. Conversion to other combinations, such as cutting a single ace of hearts shuffled into two decks, is about a 1% chance (52 cards + 52 cards = 104 cards, giving a chance of about 1 in 100 to select any single unduplicated card).

In multifactorial disorders, unlike single gene transmission, the more family individuals that are affected, the greater the risk for others to have the disorder also. Multiple affected people in a family indicate that more factors are at work to produce the disorder—genes, environment, or both. Thus, the risk is higher.

It is very important to avoid the "gambler's fallacy"—that is, since the expected event has already

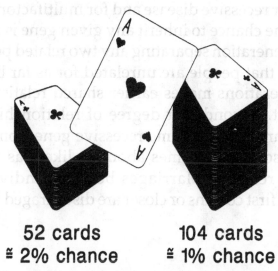

52 cards ≅ 2% chance   104 cards ≅ 1% chance

Fig. 10. Cutting decks of playing cards. There is 1 chance in 52 of selecting any specified card in a single deck. This is about the 2% risk for some multifactorial disorders.

happened, then it will not happen the next time. Not so; we have not removed the heads or tails from the coins or the ace of hearts from our deck of cards for the next try. When we flip coins, what has happened before has no influence on the next toss. Having had a child affected by CF when both parents are carriers, does not mean that the next three children will not be affected. The risk for an affected child remains 1 in 4 at each following pregnancy.

Consanguineous is a big word for describing two people who have a common ancestor, "blood relatives" by reason of inheritance rather than marriage. The problem with parents having a common relative is the increased risk for having the same abnormal gene inherited from the same source. The increased risk is mainly for recessive disease and for multifactorial disorders. The chance to inherit any given gene is halved by each generation separating any two related persons. Knowing that people are unrelated for as far back as three generations makes earlier shared relatives less important. Beyond that degree of relationship, just being a carrier of the same recessive gene from independent sources becomes about as likely as from a common relative. Marriages between individuals related as first cousins or closer are discouraged by law.

CHAPTER 2

Genetic Counseling

DEFINITION

It was not so long ago that the specialty of medical genetics had genetic counseling defined for it by a committee of the American Society of Human Genetics. That definition was published in 1974 as follows:

Genetic counseling is a communication process which deals with the human problems associated with the occurrence, or the risk of occurrence, of a genetic disorder in a family. This process involves an attempt by one or more appropriately trained persons to help the individual or family (1) comprehend the medical facts, including the diagnosis, the probable course of the disorder, and the available management; (2) appreciate the way heredity contributes to the disorder, and the risk of recurrence in specified relatives; (3) understand the options for dealing with the risk of recurrence; (4) choose the course of action which seems appropriate to them in view of their risk and their family goals and act in accordance with that decision; and (5) make the best possible adjustment to the disorder in an affected family member and/or to the risk of recurrence of that disorder.*

*From: Fraser, F. C.: Genetic counseling. *American Journal of Human Genetics* **26:** 636–661, 1974.

21

Note that the emphasis is on understanding of the overall situation and dealing with that situation as best one can. Help by suitably trained persons is a key factor. It is not the purpose of this book to substitute for help by trained persons, but only to serve to reinforce facts and suggest suitable counselors.

One of my coworkers in medical genetics has said that everybody receives genetic counseling. It may be by family members, friends, or even the corner grocer. Obvious physical deformity in a child is apt to prompt a response such as "That mother does not need to have any more children with a child like that." The opinion of the untrained is likely to be wrong based on faulty knowledge both of the cause of disease and the mechanics of inheritance. Most people on receiving proper counseling are surprised to learn that the real risk for an abnormal event to happen again is almost always less than they had feared.

WHO NEEDS COUNSELING?

Anyone who has a concern about inheritance of a disorder needs information to answer that question. Many times, the details needed can be obtained by asking other family members. Given the information on the family, the family physician or obstetrician can often provide an answer. Specialists in medical genetics are readily available by telephone to practicing physicians, and can help decide if additional laboratory testing or consultation is needed. Genetic counselors are available as part of the medical genetics team in larger medical centers. They can provide personalized help for people in need of understanding genetic disorders and in coming to grips with their feelings.

WHAT THE COUNSELOR NEEDS

Any attempt to answer questions on risks for a disorder at any level of expertise is first going to require a diagnosis. What is it that we are talking about? The family physician is often well aware of the diagnosis in question. If not, it may be readily available from other sources—other physicians, clinics, or hospital records. That diagnosis may not be an inherited disorder, and the concern can be dismissed. On the other hand, what may appear to be a sufficient diagnosis can be inadequate. For example, there are at least ten different types of muscular dystrophy. Although the most common, Duchenne or pseudohypertrophic muscular dystrophy, is X-linked, other types of muscular dystrophy are inherited in an autosomal dominant or recessive fashion. Precise diagnosis is critical as a basis for accurate counseling. Attempts to arrive at the proper diagnosis may be the largest part of the problem.

If the disorder is inherited, the family relationship of the person or persons with the disorder to the person concerned then becomes the next needed information. Descriptive terms for family kinship often become confusing once the described relationship goes beyond first cousins. First cousins once removed or second cousins may not be the correct description for a distant relative. Even at the first cousin level, there may be a vast difference in whether the relationship is through a mother or a father, especially in considering X-linked disorders. Construction of a pedigree is the accurate way to describe kinships, and, fortunately, the pedigree chart is the simplest way

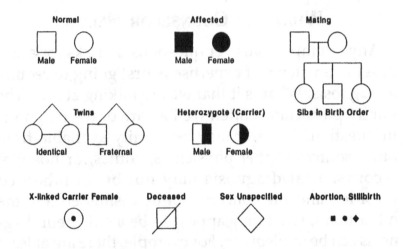

Fig. 11. Pedigree symbols. Circles for females and squares for males are connected to show marriages and children.

also. Asking about diseases and causes of death in the extended family can provide important data in a pedigree.

CONSTRUCTING A PEDIGREE

Figure 11 illustrates standard pedigree symbols. Squares represent males and circles females. Horizontal lines connect the symbols for the two parents. Brothers and sisters are referred to as siblings or sibs, and a set of brothers and sisters as a sibship. A vertical line from the parental marriage line connects to a horizontal line to which squares and circles for their children are connected to form a sibship. Symbols between a quarter of an inch and a half inch in size will usually allow a convenient size for the addition of names, birth dates, and other information for a small family. Smaller symbols are used to indicate abortions (miscarriages) and stillbirths. Taping pages together in an accordian fashion is a simple way to record an extended pedigree (a kindred). Nowadays, computer paper is handy for the purpose. The original rough draft of a large kindred I prepared consisted of 18 sheets of paper taped together. I challenged my children to count the people (750 or so), but they were never successful in arriving at an exact count. Fortunately, most pedigrees need not be nearly so large to include the branches of the family that show the affected persons. For medical genetic purposes, the idea is to trace disease genes and family relationships rather than to trace back ancestry to an important person or event, which is often the purpose in genealogy.

Sibs ideally should be listed in birth order from left to right. Darkened symbols are used for individuals with the disorder of interest, and open symbols for those not affected. Notes and question marks are often helpful. It is nice to be neat, but it is more important to be accurate. Start with the youngest generation and work back. Otherwise, there may not be enough space left between symbols in older sibships to allow for all their descendants. Grandmothers and great aunts are often helpful in producing an accurate pedigree. Medical professionals provide help in diagnosis.

A geneticist can often readily analyze a roughly drawn pedigree at a glance. This is because the pattern of affected persons in a pedigree fits one of the usual types of inheritance. Dominant disorders, for example, show vertical patterns with direct transmission from parent to child. Recessive disorders involving multiple brothers and sisters have a horizontal

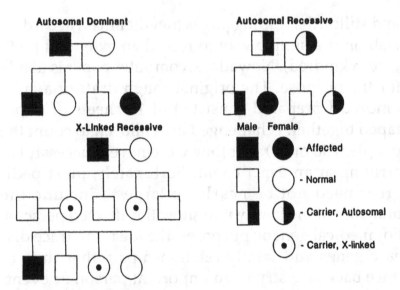

Fig. 12. Stylized pedigrees showing patterns of inheritance. The pedigrees illustrate the results of expected gene distribution for autosomal dominant, autosomal recessive, and X-linked recessive disorders.

pattern. This is the direct result of the inheritance of individual genes we have previously discussed. Stylized pedigrees depicting classic inheritance patterns are illustrated in Fig. 12.

CHAPTER 3

Prenatal Diagnosis

INDICATIONS

Prenatal diagnosis is generally used to seek an abnormality in a fetus when there is a greater than usual risk for a specific disorder. The risk must be for something that is detectable with available tests. This means that not all genetic disorders can be diagnosed. Concern for risk to fetus and mother, capability and availability of needed tests, and cost preclude routine testing for all pregnancies. As in all medical procedures, the benefit to be gained must be weighed against the hazards of the testing itself. Whether or not to undergo testing is a decision for the parents to make with the help of the obstetrician, and sometimes a medical geneticist and genetic counselor.

Advanced maternal age (35 years or greater at delivery) is the most common reason for using prenatal diagnosis. This is because chromosome abnormalities, such as Down syndrome, are greater at older maternal ages (*see* Table 1). Previous birth of a child with a chromosome abnormality or a parent with a chromosome rearrangement, is also a reason to obtain chromosome studies on the fetus. Increased risk for a neural tube defect or other physical abnormality may

Table 1
Rounded Risk for Chromosome Abnormalities
Among Live Births by Maternal Age

Age	Down syndrome	All types
15	1/1000	1/450
20	1/1600	1/525
25	1/1250	1/475
30	1/900	1/380
35	1/380	1/180
36	1/290	1/150
37	1/230	1/125
38	1/180	1/105
39	1/140	1/80
40	1/110	1/65
45	1/30	1/20

call for ultrasound study. An increasing number of recessive disorders may be detected by biochemical tests or recombinant DNA techniques for known carrier parents.

SONOGRAPHY

The use of echoes from transmitted sound employed to locate submarines has been refined for medical applications. By using frequencies very much higher than the human ear can hear (ultrasound) and small hand-held sending and receiving devices (transducers) coupled to computers, rather exact pictures of internal body parts can be obtained. Sonography is very good for showing the fetus floating in amniotic fluid, because it works well in locating differences in echoes where solids and fluids touch. It is harmless to the fetus and completely painless to the mother. The transducer is placed on the mother's abdomen and moved about while the picture is observed on a video screen and stored in the computer.

Physical abnormalities, such as the neural tube defects—spina bifida (open spine) and anencephaly (failure of the brain and top of the skull to develop)— can usually be detected with sonography. Abnormal limbs, some types of bone abnormalities, dwarfism, and even internal anomalies, such as absent or cystic kidneys and some types of congenital heart defects, may be seen under ideal conditions. One of the most common uses of sonography is to date the age of a fetus accurately by measuring the diameter of the skull. This can be done with most sonographic equipment. A careful spine examination, though, may require a more sophisticated device and a great deal more time to obtain a good picture of small internal abnormalities in a fetus free to move about. Figure 13 shows pictures of fetal nose, lips, and an ear.

AMNIOCENTESIS

Amniocentesis is the process of removal of a small portion of the fluid surrounding the fetus. It is usually performed in an outpatient clinic by the obstetrician after sonography to locate the placenta. A thin needle of similar diameter to that used to draw blood from a vein is inserted into the abdomen, but in this case, the needle must be long enough to reach through the abdominal wall and into the fluid inside the womb (Fig. 14). The needle has a fine wire inside to stiffen it and to prevent picking up cells from maternal tissues. The needle stick in the skin produces the only discom-

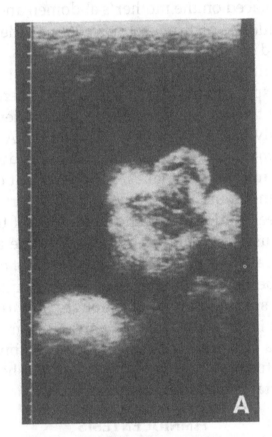

Fig. 13. **(A)** Ultrasound picture of fetal nose and lips.

fort, and this is sometimes lessened by a drop of local anesthetic injected into the skin using a very fine needle.

For genetic study, less than an ounce (20–30 cc) of fluid is removed with a syringe. At 16 weeks of gestation, the amount of fluid is great enough to provide for easy needle insertion, and the small portion removed is quickly replaced by the fetus. The fluid contains cells of fetal origin, most of which are from

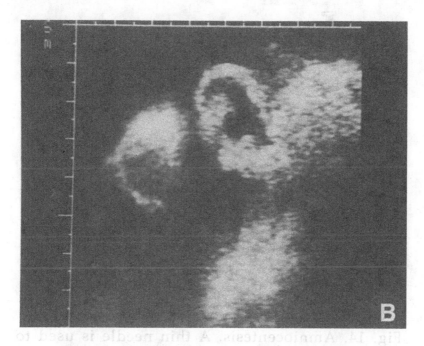

Fig. 13. **(B)** Ultrasound picture of fetal ear.

the inner layer (the amnion) of the tissues surrounding the fluid cavity. Special tests may then be performed on cells and fluid as necessary.

CHORIONIC VILLUS SAMPLING

Chorionic villus sampling (CVS) is the removal of tiny bits of tissue (villi) from the fetally derived supporting tissue (chorion frondosum) destined to become the placenta within the uterus (womb). It is generally performed between the 9th and 11th weeks of pregnancy. Using ultrasound guidance, a thin plastic tube with a stiffening wire inside is inserted through the womb outlet (the cervix) at the top of the vagina

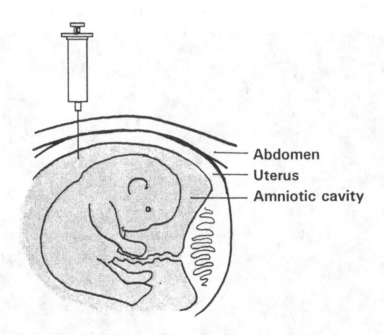

Fig. 14. Amniocentesis. A thin needle is used to remove a small sample of the amniotic fluid surrounding the fetus.

(Fig. 15). Another method uses a needle inserted into the abdomen like an amniocentesis to aspirate small bits of tissue.

A relatively smaller number of centers provide CVS in addition to amniocentesis. The very slight risk of complications for mother and fetus from CVS is slightly higher than for amniocentesis. Technical error is higher for CVS, and neural tube defects cannot be detected by it. There is probably a higher risk for Rh sensitization by CVS, requiring use of RhoGAM injections for Rh-negative mothers. The advantages of CVS are that it can be done earlier and the rapidly growing tissue obtained provides quicker results for chromo-

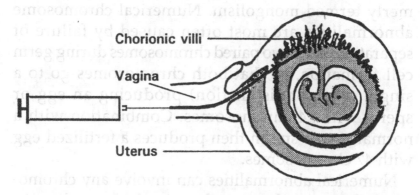

Fig. 15. Chorionic villus sampling (CVS). A small caliber plastic tube is inserted by way of the vagina through the mouth (cervix) of the womb (uterus) to remove a small sample of the chorion, which is of fetal origin.

some and biochemical tests. Thus, if pregnancy termination is necessary, it can be performed safely and inexpensively as an outpatient procedure by 12 weeks of pregnancy.

CHROMOSOME ABNORMALITIES

There are two major types of chromosome abnormalities: (1) numerical—too many or too few, and (2) structural—alteration of a chromosome as a result of breakage, sometimes with gain or loss of genetic material.

About 1 in 166 newborns has a chromosome abnormality. The most common chromosome abnormality among liveborns is Trisomy 21. Trisomy means there are three chromosomes where there should only be a pair owing to having an extra chromosome, making a total of 47 instead of the normal 46. Having an extra number 21 (*see* Fig. 1) produces Down syndrome, formerly termed mongolism. Numerical chromosome abnormalities are most often caused by failure of separation of the two paired chromosomes during germ cell formation, so that both chromosomes go to a single cell (nondisjunction) producing an egg or sperm with 24 chromosomes. Combination with a normal egg or sperm then produces a fertilized egg with 47 chromosomes.

Numerical abnormalities can involve any chromosome. Down syndrome is the most frequent chromosome abnormality seen in liveborns. This is because chromosome 21 apparently contains less genetic information than any other autosome (nonsex chromosome). Thus, Trisomy 21 upsets development less than trisomy for larger chromosomes, which frequently cause miscarriage.

Individuals with Down syndrome often have round heads, upward slanting eyes (hence the earlier term mongolism), flat nasal bridge, skin folds at the inner eyelid (epicanthus), short broad hands with a single crease in the palm, short incurved fifth fingers, lax muscles, open mouth with protruding tongue, and congenital heart defects. There are a number of other features, such that an examination by a physician accustomed to seeing patients with Down syndrome can usually provide a rather firm clinical diagnosis. Confirmation by chromosome analysis provides additional information on the possibility of mosaicism (a mixture of normal and 21 trisomic cells) and translocation (breakage and rejoining of two chromosomes).

It is important to know about mosaicism for predicting development of the affected person. Translocations increase the chance that it may happen again. Almost all persons with Down syndrome are mentally retarded to some extent. Most learn to care for their own daily needs and to perform simple repetitive tasks. Some function well enough to go to a regular school, but very few are able to make their way unassisted in our modern complex society. Congenital heart defects and special liability to infection are the major life-threatening features.

Trisomy 13 and Trisomy 18 occur in liveborns, but fortunately are far less frequent. Both have much more handicapping physical and mental defects, and rarely survive longer than a few months. Almost all infants with Trisomy 18, and most of those with Trisomy 13 have congenital heart defects.

In Trisomy 13, there are often cleft lip, small or missing eyes, extra fingers, congenital heart defects, and other features that allow ready recognition.

The physical appearance in Trisomy 18 may be rather subtle, with the most likely features being poor feeding with failure to thrive, low-set malformed ears, overlapping clenched fingers, undescended testes in males, and a short jaw.

Abnormal sex chromosome numbers also occur in liveborns. The most common is a male with Klinefelter syndrome resulting from an extra X chromosome or chromosomes, usually 47 total including 1 Y and 2 X chromosomes. Rarely there may be as many as 4 X chromosomes and a total of 49. Testis development and function are decreased, causing no sperm production and immaturity of features associated with maleness, such as beard growth and deep voice.

Females may have only 45 chromosomes, including a single X chromosome producing Turner syndrome. In patients with Turner syndrome, there is short stature, abnormal ovaries causing sterility, failure to menstruate, and decreased secondary female characteristics, such as breast development.

There is great variation in both chromosome constitution and effects on the individual in the examples noted above, many of which are beyond the scope of this volume. One variant called mosaicism certainly deserves mention. Mosaicism in its broadest genetic definition means there are some cells genetically different from other cells in the same individual. Chromosomal mosaicism is common, in which there are cells with different chromosome numbers. Turner syndrome is a good example, since some individuals having mostly cells with a single X chromosome may have a small proportion of cells containing a Y chromosome. In this instance, there is an increased risk for cancer in the ovaries requiring their removal, whereas in the usual type, removal is not necessary. There are rare cases of mosaic Down syndrome where there may be a mixture of 21 trisomic cells with enough normal cells to lessen the retardation.

Structural chromosome abnormalities include loss (deletion) or gain (duplication) of parts of chromosomes resulting from breakage and rejoining. Translocation is the breakage and incorrect rejoining of two chromosomes so that parts of each are exchanged. A "balanced" translocation consists of an exchange of genetic material in which all of it is preserved—it has simply changed position—and usually causes no problem for the person in whom it first occurs, but in the formation of germ cells in that individual, the pairing of homologous chromosomes and exchange of material may lead to gains or losses that can produce minimal to severe problems in a child. Translocation between chromosomes 14 and 21, for example, can be a cause for a greatly increased risk to have children with Down syndrome. A normal person who carries a translocation consisting of both chromosomes 21 will produce only Down syndrome liveborns. This is because a carrier's germ cells will have either two chromosomes 21 or none. In combination with a single 21 from the other parent, there will be a total of three (trisomy) causing Down syndrome, or only one (monosomy), which is lethal. This is one of the very rare instances in which a genetic abnormality is transmitted to all living children.

The structural abnormality called a ring chromosome is formed by breakage at both ends of a chromosome. The broken spots join to form a circle (ring), and the broken off tips are lost. In a similar fashion, two breaks in a single chromosome may rejoin incorrectly so that a portion of the gene order is turned around (inversion). Though the genes are all still present, germ cell formation may be upset. Since structural abnormalities may involve any chromosome, the possibilities are limitless. Fortunately, structural abnormalities are less common than numerical, happening in about 1 in 500 newborns. Structural abnormalities have been found to be the cause of repeated pregnancy loss for less than 10% of couples with two or more spontaneous abortions (miscarriages). Multiple spontaneous abortions are an indication for chromosome studies for couples when there is no other obvious explanation.

Certain places on chromosomes are especially liable
to undergo breakage, termed fragile sites. One fragile site near the tip of the long arm of the X chromosome, termed the fragile-X chromosome, is closely
associated with a gene causing mental retardation.
Using special techniques, a chromosome break or a
"pinched" area is seen on the X chromosome in up to
half of the cells of affected males, but in fewer cells in
carrier females.

The fragile-X chromosome is a rather frequent cause
of mental retardation. Affected males often have a
prominent jaw and enlarged testicles. Not all males
who transmit the fragile-X chromosome are retarded,
and some carrier females may be retarded. Transmission of the retardation to children of individuals with
the fragile-X chromosome is more complex than simple
X-linked inheritance. One possible explanation for
mentally normal fragile-X males who have retarded
grandchildren is "imprinting," in this case, the blocking of reactivation of an inactive X chromosome prior
to germ cell formation in females. Imprinting is a
reversible change in gene action rather than a change
in the DNA code itself that may be transmitted from
one generation to another. It occurs during germ cell
formation and allows different expression of DNA
depending on its inheritance from mother or father.
The cause of the fragile-X syndrome has now been
shown to be the result of changes in size of a repeated
three-letter code (trinucleotide repeat) in DNA, further
explained in the section on Special Mechanisms of
Inheritance.

BIOCHEMICAL DISORDERS

Disorders owing to abnormal single genes inherited in an autosomal dominant, autosomal recessive, or X-linked fashion as previously described may result in deficiency of specific chemical products detectable by laboratory testing. The genetic mechanisms of the hemoglobin abnormalities, such as sickle-cell disease and the thalassemias (Mediterranean anemia or Cooley's anemia), are known down to the DNA code changes in the respective genes. Other disorders either cause failure to produce or overproduction of substances that can be tested for. The routine detection of phenylketonuria (PKU), for example, depends on finding an increased amount of the amino acid phenylalanine, a normal nutrient and building block for protein. Greater amounts appear in the blood owing to defective activity of the enzyme (phenylalanine hydroxylase) that converts phenylalanine to another amino acid, tyrosine. By careful control of the diet, the devastating associated mental retardation apparently can be reduced in most cases.

For disorders as common as PKU and hypothyroidism, screening of the entire population of newborn infants has proved to be practical. Other disorders, such as galactosemia, have been added to some newborn screening procedures, but so far, it is not practical to screen for all potential disorders.

Testing of parents in high-risk groups for the carrier state has been another strategy for the prevention of disease in their children. Identifying African-Americans who are sickle-cell carriers and Ashkenazi Jews who are Tay-Sachs carriers is helpful. Recent discovery of the most common mutation causing CF was followed by discovery of more than 20 less common mutations. Using current technology, only about three-fourths of the couples at risk for having a child with CF would be detected, so that some feel that mass screening must wait. Meanwhile, testing is recommended only for couples with a family history of CF.

The process of mass screening and counseling has posed multiple problems. Dealing with large populations rather than individuals makes things difficult, such as keeping records private, not to mention the task of providing counseling to many, many people.

Other disorders producing enzyme deficiencies are often detected by diagnosing the disorder in an affected child. Parental carrier states are then determined by pedigree analysis according to known inheritance patterns (most are autosomal recessive). In some instances, the carrier state is confirmed by enzyme tests on the parents. In many cases, cells obtained by amniocentesis can be used to test for the abnormality in subsequent pregnancies. For this strategy to be of value, it must be known that parents are carriers of a disorder that can be detected using cells or fluid obtained prenatally.

α-Fetoprotein

α-Fetoprotein (AFP) is a protein that is produced by the growing fetus and reaches its greatest concentration in fetal blood at 14–16 weeks of gestation. Since it is normally excreted by the fetus into amniotic fluid in small amounts, AFP measurement in amniotic fluid has proven to be a reliable test for neural tube defects.

Neural tube defects are the result of failures in formation of the central nervous system. In anencephaly, the higher brain centers and the top of the skull do not develop. In spina bifida, the defect is in the bone, muscle, and skin covering the spinal cord. Anencephalics rarely survive longer than a few hours, and individuals with spina bifida are usually severely handicapped.

Much larger amounts of AFP leak into amniotic fluid through the thin membranes covering neural tube defects. Testing for AFP in amniotic fluid allows their detection. AFP derived from the fetus also enters the mother's bloodstream, but in much smaller amounts. Using very sensitive tests, though, even increases in the tiny amounts of AFP in maternal blood in cases of neural tube defects can be detected, although not with the same accuracy.

Testing of maternal serum AFP, often referred to as MSAFP, is a reliable screening test. Problems of falsely high or low results occur. When MSAFP is increased, ultrasound and amniotic fluid AFP testing usually are performed to seek a definite answer about a possible neural tube defect. Such factors as age of the fetus, race, and weight of the mother influence MSAFP testing. Fetal abnormalities other than neural tube defects can cause increased amounts of AFP. Kidney disease, twin pregnancy, and hernias into the umbilical cord (omphalocele) are some of the things that can cause elevated AFP. Having a decreased amount of MSAFP is associated with an increase in risk for Down syndrome, though the reason for this finding is not fully understood.

Use of a combination of tests for AFP and pregnancy hormones helps to increase the accuracy of prediction of fetal abnormality.

CHAPTER 4

Teratogens

Teratogens are environmental agents that can cause birth defects, or affect development of an embryo or fetus directly through exposure of the mother during pregnancy. The action does not occur by altering the genetic makeup of the embryo, although some agents, such as X-ray, can act as both a teratogenic agent and a mutagen (changing a gene by altering the DNA code).

The timing of exposure of the embryo or fetus is critical in causing a defect. The most sensitive period is from the 18–40th day after conception when the organs are being formed.

Prescription medications are tested for harmful effects in pregnant laboratory animals, but in many cases, there has simply been too little experience in humans to be certain that a drug is safe in pregnancy. Thus, the wise attitude is to take no medication other than what is necessary during a pregnancy.

For all known teratogens, there is a dose effect. That is, exposure below a certain amount causes no damage. Some agents that can cause birth defects are listed in Table 2.

43

Table 2
Some Agents That Can Cause Birth Defects

Alcohol	Lithium
Androgenic steroids	Maternal diabetes
Anticoagulants (some)	Maternal PKU
Anticonvulsants	Mercury, organic
Cancer chemotherapy	Parvovirus
Cocaine	Penicillamine
Cytomegalovirus	Radiation
Estrogenic steroids	Rubella
Etretinate	Streptomycin
Herpes virus	Syphilis
Inorganic iodide	Tetracycline
Isotretinoin	Thalidomide
(Accutane)	Toxoplasmosis

Some things that have been suspected to cause birth defects, but probably do not, include: aspirin, birth control pills, spermicides, antinauseants, some illicit drugs (marijuana, LSD), video display terminals, anesthetics, aspartame, rubella vaccine, metronidazole, and agent orange.

Women who drink alcohol daily or have binge episodes of alcoholism during pregnancy may have a baby with the fetal alcohol syndrome. Fetal alcohol syndrome consists of poor growth, small head, abnormal facial features, other internal abnormalities, and learning disability in childhood or mental retardation. Although an occasional drink may be safe, using no alcohol is safer, since it is not known how little alcohol can cause problems.

Cocaine abuse during pregnancy has only recently been recognized to cause problems in fetal development. Prematurity, growth retardation, and small heads have been reported following cocaine use. Blood vessel abnormalities in the brain leading to death of brain tissue, and abnormal urinary or sex organs have been seen following cocaine use, though a clear pattern such as in fetal alcohol syndrome has yet to be fully defined.

Avoidance of any drug or medication except those essential for treatment or health maintenance has to be the best answer during pregnancy.

Chapter 5

The New Genetics

RFLPs, PCR, AND DNA FINGERPRINTS

The discovery of easily testing for normal differences in the DNA code among individuals has provided a powerful new tool for genetics. Using the principles of genetic linkage (page 11), disease genes about which rather little is known can often be traced. RFLP is an acronym using the first letters of the complex sounding term: restriction fragment length polymorphism.

A polymorphism is the occurrence of two genes in a population with the least frequent gene found in at least 1% of the population. This provides a reasonable chance to find the rarer gene.

Restriction refers to an enzyme that will cut DNA into pieces by attacking it at places where there is a specific sequence of the DNA code. A couple of hundred such enzymes called restriction endonucleases have been discovered, mostly in bacteria. These enzymes chop DNA into many small pieces, since the short recognition code occurs by chance many times in the complete gene set. The presence or absence of the specific short DNA coding sequence matching a specific restriction enzyme allows cutting only at those sites to produce different lengths of DNA pieces. These DNA fragments are separated using an electric current (electrophoresis) and then transferred to a thin membrane for easier handling. Detection of the sizes of the DNA pieces using probes can then identify the presence of the sequence in one or both of a person's chromosomes.

The probes are segments of DNA that have been made radioactive and have a sequence that allows them to combine only with fragments on the membrane with the same code. The radioactivity is then used to produce a spot on photographic film.

When looking for polymorphic restriction sites in DNA that correlate with the presence or absence of a genetic disease, the site is used as a linkage marker locus. Finding a restriction site close to a disease gene locus can then be used to track the disease gene. This strategy works without knowing what the DNA function is at the site—only that it is close to the disease gene locus. There is still a possibility of exchange of parts of chromosome pairs between the two loci during germ cell formation. Therefore, the restriction site must be very close to the disease gene to be of practical value in prenatal diagnosis or carrier detection. In some cases, though, there are restriction sites that recognize the code for the genetic abnormality—the actual DNA mutation producing the disease. This is the case for sickle-cell hemoglobin. Detection of the abnormal gene itself is provided rather than a linkage marker close to the gene.

In addition to sickle-cell disease, the RFLP technique has already been successful in a growing number of disorders, such as thalassemia, CF, Huntington's disease, polycystic kidney disease, hemophilia, and X-linked muscular dystrophy (Duchenne and Becker types) to name a few.

PCR, which stands for polymerase chain reaction, is a new system that greatly expands the usefulness of RFLPs and has other applications as well. PCR is basically a method for rapidly copying segments of DNA in the test tube. It requires knowledge of the sequence of a short segment of the DNA code on either side of a longer segment of interest. The known sequence is chemically synthesized—a relatively simple process automated in a "gene machine." The two single-stranded short sequences are then allowed to bind to the two opposite strands of DNA to be copied in a test tube and serve as starting points for DNA duplication. A mixture containing all the chemical building blocks and enzymes for DNA synthesis allows repeated copying of the DNA segment between the two short beginning sequences. The reaction is controlled by temperature, cycling the separation of the two complementary DNA strands, and growing new strands on each old strand. This process can select out a single DNA segment of interest even in a mixture of the whole human genome because the small starting sequences, or primers, recognize and bind only to the specific complementary DNA code. PCR can produce millions of copies in a few hours. Large enough amounts of material often are obtained to use chemical stains not requiring radioactive materials or the longer time periods needed to detect radioactivity. This gives quick results using tiny amounts of material. Use of the PCR

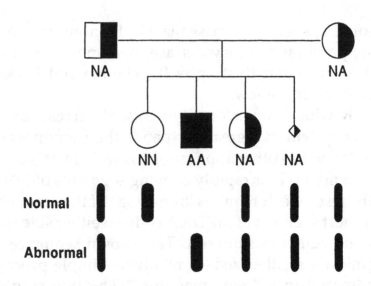

Fig. 16. Stylized pedigree showing diagnosis of CF using PCR (below). Normal, affected, and carrier children, and a carrier fetus are included.

technique in diagnosing cystic fibrosis is illustrated in Fig. 16. The stylized pedigree shows what would be expected by amplifying the CF gene and the normal gene using primers specific for each. The result shows the presence of both genes in carriers, and only one type of gene or the other for normal or affected. Some CF genes cannot be detected because they are different from the common mutation shown.

Some DNA sequences repeat many times in a chain. The number of repeats vary such that they provide a unique pattern for an individual. The patterns of DNA fragments may be very complex, depending on the particular method and probes used. These methods produce "DNA fingerprints," which can be used to determine genetic identity of cell samples. The tech-

Table 3
Frequency of Recessively Inherited Disorders

Disorder	Incidence at birth, 1 in	Carrier frequency, 1 in
CF	2000	23
Phenylketonuria	15,000	62
Sickle-cell disease	400	10
Tay-Sachs disease	4500	34

nique is already finding application in medicolegal matters, such as identifying parents or evidence obtained at the scene of a crime.

Some radioactive or fluorescent probes can be used to locate DNA sites on chromosomes directly in the microscope. These are termed *in situ* techniques.

SOME COMMON INHERITED DISEASES

Nothing is so common as the rare disease to the person who has it. Thus, "common" may be quite different depending on who is looking at the problem.

Two of the more common single-gene inherited diseases in the United States have already been mentioned: CF among Caucasians and sickle-cell disease among African-Americans. Both are inherited in an autosomal recessive fashion. The carrier rate for recessive disease is much higher than the frequency of affected persons (page 15). One view is to look at recessively inherited disorders from the standpoint of both frequencies in Table 3.

Carrier frequencies for autosomal recessive disorders can be estimated from the number of affected people in a population using the Hardy-Weinberg equation.

For rare disorders, carriers occur in a frequency roughly two times the square root of the frequency of affecteds. For normal brothers and sisters of affected persons, there is a 2/3 risk for being a carrier. One normal:two carrier:one affected children of two carriers yields two carriers out of three unaffected children (Fig. 8A, page 15). The carrier risk for an apparently normal sibling of an affected person married to a nonrelative becomes 2/3, times the population carrier rate for the spouse, times one in four for an affected child if both parents are carriers. For CF, for example, the risk would be $2/3 \times 1/25 \times 1/4 = 1/150$, or in card-cutting terms: one ace of hearts in three decks of cards. Fortunately, there is now a specific test for the CF gene so that the carrier state can be known rather than estimated. Gene frequency estimates are still helpful in many other disorders for which there is no specific carrier test. For rare disorders at frequencies of 1 in 40,000 or lower, the carrier frequency is less than 1 in 100.

Many inherited disorders are caused by lack of activity of a specific enzyme. Enzymes are proteins that change one substance to another in a single chemical step. Enzyme defects are often inherited in an autosomal recessive fashion. A gene on one chromosome producing a bad enzyme usually does not produce a severe deficiency. This is because the normal gene at that locus on the other chromosome provides enough enzyme, but if both genes are abnormal, there is likely to be too little enzyme function. Such enzyme defects have been termed metabolic disorders, because the enzymes are involved in the body chemistry.

Problems arising from these disorders generally result from a lack of some necessary substance that should be produced or excess of a substance that would ordinarily be converted to another. More or less effective treatments have been devised depending on the defect. Dietary restrictions reducing intake of phenylalanine in phenylketonuria, other specific amino acids in certain aminoacidopathies, galactose in galactosemia, and lactose in lactase deficiency have proven useful. Replacement of deficient products is used, such as pancreatic enzymes in CF, cortisone in adrenogenital syndromes, and thyroid hormone in congenital goiter. Removal of excessive amounts of stored material, such as copper in Wilson disease, iron in hemochromatosis, and uric acid in gout, may be helpful.

Many common diseases are of multifactorial origin, with genetic factors and environmental factors combining to produce disease. Birth defects, such as heart malformations, the neural tube defects (anencephaly and spina bifida), and cleft lip, are examples. Although the majority of persons suffering from such disorders have them on a multifactorial basis, there are some exceptions in which single genes produce the disorder. One type of cleft lip is the result of a single dominant gene and is often separable from the more common multifactorial cause by finding small pits or cysts on the lower lip. There are many syndromes in which cleft lip may simply be one part of the broader syndrome. Diabetes mellitus, epileptic seizures, and mental retardation are usually multifactorial, but may also in certain instances be the result of the effect of genes at a single locus. There are probably many different genes at different loci on the human X chromosome that can cause mental retardation.

One of the major differences between multifactorial and single-gene causation is that the recurrence risks are much less for multifactorial disorders. The usual risk for a child to be affected with a multifactorial disorder is about 2–5% when there is only a single affected first-degree relative (parent, brother, sister, or child).

CHOLESTEROL, HEARTS, AND HEADS

Cholesterol is an essential part of cell membranes, steroid hormones, and bile acids sent from the liver into the bowel for food digestion. It will not dissolve in water (insoluble). This makes cholesterol an important part of cell membranes, but its insolubility is also what makes cholesterol lethal when it builds up in the wrong place. Storage of cholesterol in the wall of arteries, termed atherosclerosis, causes coronary heart disease and stroke by blocking the flow of blood to heart muscle or brain. Coronary heart disease is the leading cause of death in industrial societies, but varies greatly in frequency with geography and nutrition.

We eat cholesterol, and it is also made by the liver. The handling of cholesterol is complex. Since it is water insoluble, it is carried in the blood bound to several different proteins called apoproteins. Apoproteins attach to cell surfaces at specific sites called receptors, which allow the entry of the cholesterol/protein complex (lipoprotein) into the cell. Cells receiving too little cholesterol manufacture cholesterol themselves. This results in many places for cholesterol handling to go wrong.

Familial hypercholesterolemia (FH) is one of the most common dominantly inherited diseases. With a worldwide frequency of 1 in 500, it accounts for about 5% of heart attacks below the age of 60. People with one FH gene are born with blood cholesterol levels twice as high as normal adults, and those rare individuals (about one in a million) with two FH genes have up to six times normal cholesterol. With one FH gene, heart attacks begin occurring around age 35, and in the homozygotes with two FH genes as early as age five. FH blocks entry of cholesterol into cells by decreasing receptors, trapping it in the bloodstream, and depositing it in arteries.

Many other genes direct the structure of apoproteins or enzymes affecting the control of blood cholesterol. Genetics accounts for 50–60% of the control, and environment—in this case, mostly diet—the rest. Diet and, in some cases, medication can be used to maintain a satisfactory blood cholesterol level below 200 mg% even for those whose genetic makeup tends to cause high values.

TRANSPLANTATION

Transplantation of organs from one person, the donor, to another, the recipient, requires that both people be as much alike as possible genetically. The most important genes are those that code for the structure of the surface of cells. Certain parts of the cell surface, especially sugars attached to proteins, can act as antigens. Antigens are things seen by the recipient as foreign or "nonself" that are capable of causing antibodies to be produced by the recipient. In addition to antibodies, certain "killer cells" may be activated to attack and kill the transplanted tissue or organ. The antibody system is one of the most important protective mechanisms for us in fighting bacteria and viruses. It is the basis for immunization against tetanus and polio, for example, but when a kidney, heart, or bone marrow is transplanted, such immune reaction is undesirable.

Most people are aware of blood group substances on the surface of red blood cells that specify whether we have A, B, AB, or O blood type and are Rh-positive or Rh-negative. These substances are of greatest importance in transfusing blood, but there are even more complex substances on the surface of nucleated cells in tissues (a red blood cell has no nucleus). There is a series of genes lined up on chromosome 6 called the major histocompatibility complex (MHC) or simply the HLA region. There are separate genes at five loci, with an enormous number of possibilities. The HLA-A locus has at least 20 different genes, and the B locus has more than 40. There is an additional series of genes at the C, D, and DR loci. Since there are so many possibilites, it is very difficult to find two unrelated people who have exactly the same tissue type. On the other hand, brothers and sisters have a one in four chance of having received the same HLA types from their parents. Identical twins have exactly the same type since they are from a single fertilized egg that has split to form two individuals. This is why related donors are so much more likely to have a "match" for transplantation.

Fortunately, there are medicines that overcome part of the rejection problem, so that transplantation between individuals who do not have an exact match can succeed.

The great genetic variety at the HLA loci also makes the tissue typing process very useful in checking parentage. A man wrongly accused as the father of a child can often be cleared. Likewise, the odds may be found in some cases to be very likely (a million to one or so) that a person is the father of a child, or in other cases the mother, when there has been a possible mix-up of babies in a nursery.

CANCER GENETICS

Cancer, as the second leading cause of death in the United States, is sandwiched right in between coronary heart disease and stroke. Some specific cancers, such as retinoblastoma, a malignancy of the retina of the eye, have long been known to be inherited in some families. Another dominant gene termed familial polyposis of the colon also causes cancer. It actually causes benign tumors of the large bowel, but there are so many small tumors that one or more will almost always change into cancer by the age of 45.

The development of cancer usually involves more than one step. In retinoblastoma, an inherited abnormal gene—or in some cases, a small missing piece (deletion) of chromosome 13—may be the first event. A second event is required to produce a tumor, usually a change (mutation) in the remaining normal gene on the other chromosome.

In chronic myelogenous leukemia, there is usually a translocation (exchange of chromosome material) between the long arm of chromosome 9 and chromosome 22 in the cells of the bone marrow. This is an acquired abnormality, because the chromosomes of tissues other than that for blood cell formation in the bone marrow are normal. There are other chromosome changes found in other leukemias and cancers, but the changes are not as exact as in chronic myelogenous leukemia. In some cases, the point of breakage on the chromosomes has been found to be at the site of genes called "oncogenes," resulting in change of the DNA coding.

Oncogenes were first discovered in certain viruses that could produce tumors in animals. Later it was found that the viral oncogenes were actually present as normal genes that had been altered in the process of transfer from animal to virus. A number of such genes have been found on many different chromosomes. The normal gene from which the viral oncogene is derived is termed a proto-oncogene. Some proto-oncogenes produce proteins with enzyme activity called kinases, but the overall function of oncogenes is still incompletely understood. They may be normally active during embryologic development, but when altered, these nomal genes lose their control of cell growth, allowing the development of a tumor. Other genes called "suppressor genes" are known to curb tumor formation. The complex multistep mechanism behind the development of specific cancers has yet to be completely understood. It is certain to be different for different types of cancers and leukemias.

SPECIAL MECHANISMS OF INHERITANCE

Just as in chromosomal DNA, the DNA of mitochondria can change to cause genetic defects. As previously noted, the mitochondria are small bodies (in the cell cytoplasm) that provide energy for cells (*see* Fig. 4). They are given to all children by their mothers only. With discovery of the entire sequence of mitochondrial DNA, an increasing number of associated human disorders have been found.

A variety of disorders are the result of mitochondrial DNA mutation including some muscle and eye disorders, such as the following examples:

- Cardiomyopathy from heart muscle weakness causing heart failure;
- Sensitivity to toxicity of the antibiotics chloramphenicol and streptomycin;
- MELAS syndrome, which includes migraine headaches, hearing loss, cataracts, seizures, strokes, and muscle weakness;
- Leber optic atrophy causing visual loss;
- Leigh syndrome, showing muscle weakness, with floppy limbs, wobbly walking (ataxia), pigment changes in the retina of the eye, degeneration of parts of the brain, and mental retardation;
- Diabetes-deafness syndrome, where deafness is associated with noninsulin-dependent-type diabetes mellitus; and
- Kearns-Sayre syndrome with muscle weakess especially of the external muscles of the eye, heart muscle weakness, pigmentation and degeneration of the retina, ataxia, and small stature.

Increases in the number of repeat DNA elements are the cause of several disorders. Trinucleotide repeats (long chains of three bases, e.g., CAG,CAG,CAG, . . .) are the basis for Huntington's disease, Spinocerebellar ataxia, Myotonic dystrophy, and the Fragile-X syndrome. The repeat DNA in these disorders is "unstable" and susceptible to changes in the number of repeat sequences. The changes are influenced by the sex of the transmitting parent, producing a complex pattern of inheritance, especially as to severity and age of onset of the disorder. The exact reason similar changes in the DNA produce the specific disorders has yet to be determined.

THE HUMAN GENOME

The complete sequence (C, G, A, and T) of DNA structure in the chromosomes is termed the human genome. The entire structure of mitochondrial DNA is known, and short sequences (DNA strands) have been discovered for many individual genes. Even a rather lengthy segment—up to eight million or so DNA code letters—is known for chromosome 21, the smallest chromosome. The largest known continuous segments, though, are small by comparison to the total sequence of three billion base pairs.

Knowledge of the entire human genetic code should prove to be of immense value in detecting, preventing, and treating the more than four thousand known inherited diseases, with many applications to the cancer problem as well. One difficulty has been that the original estimate of total project cost to determine DNA sequences would be about three billion dollars. The cost has been reduced, though, by later technology.

Not all of the sequences would be for important genes, since some DNA is repetitive and may be "junk." Thus, most strategies have been devised to sequence the genes for important diseases like CF, for example, where a deletion of three base pairs coding for a single amino acid is the common mutation. Duplication of effort has resulted, but cooperation between different research groups, particularly in the search for the CF gene, has also resulted. Refinement of technology has already reduced the cost.

In addition, there have been increased manpower needs to support the task of determining the anonymous DNA sequences. The excitement of discovering the cause of specific diseases has spurred directed research behind much of the progress made so far. The genome project will undoubtedly allow additional exciting discoveries from the laboratory work as it progresses. This should help push the project to completion.

One day, we will know the entire human genetic blueprint. It should then be possible to screen a cell specimen for all inherited disease and even for genes involved in such complex disorders as alcoholism or schizophrenia. The increased means for detection will lead to increased problems in addition to providing help in treating disease. Information usage by prospective parents, individuals, insurance companies, employers, and even governments will probably lead to conflicting ideas depending on the differing goals of those who possess such information. Heated debate has resulted from screening programs in the past, and much has been learned. It is to be hoped that the social and ethical issues already opened will continue toward reasonable solutions as technical progress toward mapping the human genome continues.

CHAPTER 6

Getting Your Questions Answered

YOUR PERSONAL PHYSICIAN

Many questions about inherited diseases can be answered by anyone's personal physician. In addition, the family physician usually has the clear advantage of having information on multiple generations. His knowledge of the family pedigree is already available, although it may not have been reduced to a formal diagram. Many times, a personal physician can provide the proper advice about inherited diseases just as for other routine illness. On the other hand, many inherited disorders are rare, and there is new information about them so that help from a specialist may be needed. Most physicians have had training in basic genetics and are skilled in explaining things to patients. Thus, often all that is needed is information from a geneticist or a medical reference as to the inheritance pattern and the method of diagnosis of a specific disorder. In such cases, a telephone call or a quick look in a book may be all that a general physician needs to provide the help that is needed.

In more complex cases, the family physician may refer the patient to a medical genetics specialist, just as for complex cases in other areas of medical practice. The medical geneticist also will have reference sources for rare disorders and access to computerized databases for genetic and teratogenic information.

REFERRAL TO A GENETICIST

A medical geneticist is a person specially trained in the mechanisms of inheritance and the diagnosis and management of genetic disorders. Just as in other medical specialties, such as surgery or obstetrics and gynecology, there has been a period of special training beyond medical school. Many medical geneticists practice genetics as a subspecialty, having already specialized in pediatrics, internal medicine, or obstetrics and gynecology, for example. The American Board of Medical Genetics (ABMG) certifies physicians and Ph.D. specialists by examination in several areas—clinical genetics, clinical cytogenetics, clinical biochemical genetics, and clinical biochemical/molecular genetics. Clinical genetics deals with the diagnosis and management of patients, cytogenetics with laboratory diagnosis of chromosome abnormalities, and biochemical genetics with laboratory diagnosis and management of enzyme disorders and resulting chemical abnormalities.

Genetic counseling also was formerly certified by the ABMG. A new certification board, The American Board of Genetic Counseling (ABGC), has now taken over the growing task of certification for genetic counselors.

Most medical genetics specialists and genetic counselors are associated with large medical centers or referral laboratories for genetic testing.

WHAT TO EXPECT

Medical geneticists carry out their tasks just as most general physicians do: first gathering information, then putting that information together to make a diagnosis, and finally providing alternatives for a course of action to the person(s) consulted. In some instances, the family pedigree is the most important part of information gathering. It is thus wise for the person consulting a geneticist to arrive with as much family data as possible. Having an older member of the family available is often the most practical approach. Drawing a pedigree as previously outlined is usually a simple matter and can prove to be of great benefit. A roughly drawn family tree can often be easily interpreted by a skilled geneticist at a glance.

In some cases, cytogenetic studies—chromosome studies—may be needed. These usually require only a single test tube of blood, but as previously explained, a period of time is needed for the complex analysis to be done. For studies of amnion cells and other tissues, 10 days or longer may be needed. For many biochemical tests, urine or blood is used, and some tests can even use the tiny speck of live cells at the base of plucked hairs. Small bits of tissue obtained from skin biopsy are sometimes needed for analysis. Some of the more recently devised tests on DNA can even be performed using cells obtained by rinsing the mouth with fluid.

EARLY DETECTION AND PREVENTION

Newborn screening is now routine for the detection of such disorders as PKU, sickle-cell disease, and hypothyroidism. Finding the genetic abnormality allows treatment to be started early to prevent some of the problems of the disease. Newborn screening for disorders that can be detected at low cost and for which treatment is available is a logical approach. Opinions differ as to what diseases meet the requirements for screening. Population screening for carriers of recessive genes to identify parents who are at risk to have affected children is also effective. Such population screening uses the strategy of testing groups in which the gene is more common. Examples are testing for Tay-Sachs in Ashkenazi Jews, and for sickle-cell disease in African-Americans. Genetic counseling and prenatal diagnosis can help the carriers found by screening such groups.

TREATMENT

Some inherited disorders like diabetes mellitus are so common that we do not often even think of them as inherited. At the other extreme are genetic disorders for which there is as yet no specific treatment, such as Down syndrome. However, there are many genetic disorders for which there is treatment, even though the cause cannot be removed.

Several types of treatment are useful, as in the following examples:

Dietary limitation
  Phenylalanine in PKU
  Lactose in lactase deficiency
  Fats in hyperlipidemias

Dietary and hormone supplement
   Vitamin D and phosphate in vitamin D-dependent
   rickets
   Cortisone in congenital adrenal hyperplasia
   Thyroid hormone in congenital goiter

Removal of excess substance
   Iron in thalassemia
   Copper in Wilson disease
   Uric acid in gout

Replacement of missing substance
   Insulin in juvenile diabetes mellitus
   Factor VIII in hemophilia
   Growth hormone in pituitary dwarfism

Surgery
   Repair of congenital heart defect
   Colectomy in familial polyposis of the colon
   Splenectomy in hereditary spherocytosis

Transplantation
   Bone marrow in lysosomal storage disease
   Kidney in polycystic kidney disease
   Lung in cystic fibrosis

Gene therapy in which an abnormal gene is repaired or
replaced with a normal gene is in the experimental stage for
humans. Transfer of genes, such as for growth hormone, has
been successful in animals. Initial trials of gene transfer in
humans are in progress, and so gene therapy for human
disease seems near. Tough obstacles still remain, such as
inability to put a gene into an exact spot on a chromosome
and making the gene active in the tissues that need it.

FOUNDATIONS AND SOCIETIES

The search for diagnosis and treatment and for mutual
support has spawned societies for many genetic dis-

orders. Some are well known to the general public, such as the Cystic Fibrosis Foundation, Muscular Dystrophy Association, and the National Hemophilia Foundation. A listing of all such societies is beyond the scope of this little book, but in addition to those agencies devoted to specific diseases, there are some that provide information services to aid people in contacting the right source for help.

Some of these resources are:

Alliance of Genetics Support Groups, 35 Wisconsin Circle, Suite 440, Chevy Chase, MD 20815-7015 (301)652-5532.

American Sickle Cell Anemia Association, 10300 Carnegie Ave., Cleveland OH 44106 (216)229-8600.

March of Dimes Birth Defects Foundation, 1275 Mamaroneck Avenue, White Plains, NY 10605 (914)428-7100.

National Association for Sickle cell disease, Inc., 3345 Wilshire Blvd., Los Angeles, CA 90010 (213)736-5455.

National Ataxia Foundation, 750 Twelve Oaks Center, 15500 Wayzata Blvd., Wayzata, MN 55391 (612)473-7666.

National Cystic Fibrosis Foundation, 6931 Arlington Road, Bethesda, MD 20814 (800)344-4823.

National Down Syndrome Congress, 1605 Chantilly Dr. NE #250, Atlanta, GA 30325 (404)633-1555.

National Down Syndrome Society, 666 Broadway, New York, NY 10012 (212)460-9330.

National Gaucher Foundation, 19241 Montgomery Village Ave., Gaithersburg, MD (301)990-3800.

National Genetics Foundation, 68 Plymouth Dr., Norwood, MA 02062 (617)769-7936.

National Hemophilia Foundation, 110 Greene St., New York, NY 10012 (212)219-8180.

National Marfan Foundation, 382 Main St., Port Washington, NY 11050 (516)883-8712.

National Muscular Dystrophy Association, 3300 East Sunrise Dr., Tucson, AZ 85718 (520)529-2000.

National Neurofibromatosis Foundation, 141 5th Ave., New York, NY 10010 (212)460-8980.

National Organization for Rare Disorders, P.O. Box 8923, New Fairfield, CT 06812 (203)746-6518.

Rett Syndrome Association, Inc., 9121 Piscataway Rd., Clinton, MD 20735 (301)856-3334.

Support Organization for Trisomy 18, 13, and Related Disorders, 2982 S. Union St., Rochester, NY 14624 (716)594-4621.

Turner's Syndrome Society, 15500 Wayzata Blvd., Wayzata, MN 55391 (612)475-9944.

CHAPTER 7

Dealing with Your Decisions

NATURE: WHY DID THIS HAPPEN?

Mutations that change normal genes to new genes happen all the time. There is a risk for everyone to receive a new mutation. Fortunately, that risk is low for any single gene—about 1 in 100,000. As explained before, dominant mutations produce an immediate effect, since only one gene is required. In recessive mutations, the effect must wait until a person has inherited both genes of that mutant type at some generation in the future. Although exposure to radiation or other gene-altering agents may increase the risk for a given person or group to have a mutant gene, the overall mutation rate applies to the whole world population. Thus, the laws of probability suggest that new mutations are random events. It is like raindrops falling evenly over a roof. A good example is the calculation of life expectancy tables. For a given age in a very large population, the number of people who will die in any year can be predicted very closely from experience, but it is not known which individuals of that group will die. People who have a new mutant gene

71

have been simply unlucky in the lottery of genetics. Knowing the mechanisms of inheritance, we have the opportunity in many cases to avoid genetic disease in our children.

RISKS: WHY DID THIS HAPPEN TO ME?

Although the laws of probability tell us that genetic events are random, the risk may be vastly different given specific information. The one in a hundred thousand to have a new dominant mutation for chondrodystrophic dwarfism becomes one in two for the child of a dwarf to inherit the disorder. These principles have been explained earlier and are readily understood. Yet many people ask "why?" not in the sense of scientific understanding, but in the personal sense of "Why me—why not someone else?" There is often a feeling of guilt for a person that he or she has done something to cause the unhappy event—"Why am I being punished?" Despite concerns about drug usage, for example, by some parents of affected children, the answer is that almost always there has been nothing done by a parent to cause the problem. It has just happened by chance.

For many individuals, this explanation is enough. We cope with loss by a rather standard response: first denial, then anger, followed by understanding, grief, and, finally, acceptance and proper response. We do this all the time without wondering why, especially for simple events.

Our first response to bad news is disbelief (denial), but when we realize that the bad news is correct, we become angry. We gain understanding by thinking about what has happened and why it happened. Understanding is followed by sorrow (grief) and, after a time, by acceptance of things as they are. Finally, we continue on our way, having overcome the effects of the experience, though not having forgotten the event itself.

For trivial events, this process is rapid and complete. In serious situations, the grieving process takes longer, and loss replacement may not be possible. Some people may become stuck somewhere along the coping process and need additional help at the psychological level. Help is available both from mutual support groups and from trained professionals.

BURDEN: WHAT DOES IT MEAN TO ME?

The depth of feeling about any given situation is unique to the person involved. No two people can feel exactly the same even if they have faced what appears to be the same event. Even when that situation involves having a child with the same genetic defect, other details are always different. How bad that experience is for one person is never exactly the same as for someone else.

I have helped couples arrive at very different decisions based on the combination of burden and risk. One couple who had an infant that died of CF, having no living children, accepted the one-in-four risk of having another child affected with a lethal disease. Another couple having a living child with Down sydrome to care for was unwilling to accept the risk of 1 in 100 to have another child with Down syndrome.

Yet another couple, having lost two pregnancies to severe central nervous system defects, could not accept the increased risk (intermediate between CF and Down) for having another affected child. The important point is that only the persons involved can know what the actual—or possible—situation means to them. All of these examples occurred prior to having the current means for prenatal diagnosis, which is now available for all three examples. Many couples who wanted children in the past, but made the decision not to have them, now have additional options.

REPRODUCTIVE OPTIONS

Couples faced with an increased risk to have a child with a genetic disorder now have a number of choices. One option is simply not to have children. In our modern society, there is no longer the social stigma attached to couples being childless. Many people prefer to devote their lives to pursuits other than rearing a family. There even have been scientific studies reporting that couples without children are no less happy than couples who do have children, but for some, having children is one of their most important goals.

Adoption has been a ready alternative in the past. It remains an option, but there are fewer infants for adoption than in the past. In addition, those infants placed for adoption may themselves have increased risk for handicapping conditions. Some adoption agencies routinely ask prospective parents if they would be willing to accept an infant born to a mother who has seizures, used drugs, or a list of other possibly adverse conditions.

In some situations, artificial insemination—a pregnancy using sperm from a man other than the husband—is a suitable choice. It is most useful when the husband has a dominant disorder, but less so for recessive disorders when the sperm donor also has a risk for carrying a gene that cannot be detected.

For those couples with increased risk, some may be willing simply to accept that risk. They go ahead and have children without further consideration. Others may wish to use the option of prenatal diagnosis if it is available for the disorder in question.

As previously explained, amniocentesis or CVS for chromosome analysis or other specialized testing can in many cases determine whether or not a fetus is affected. Fortunately, in far and away the majority of cases, prenatal testing assures that the fetus does not have the abnormality. If the fetus is found to be affected, then the pregnancy can be terminated. The difficult decision for or against termination must be that which is best for the parents themselves and their family, given the available information about the prospects for the fetus. Within the bounds of sound medical practice as to timing of an abortion, no one else should make that decision for them. Most parents would abort a fetus affected with anencephaly where there is no development of the higher centers of the brain and all such affected liveborns die soon after birth. Other less severely affected fetuses require careful consideration of all the factors affecting that decision.

Life is filled with decision making. Those decisions involving having a family may be very difficult. Sometimes, especially where pregnancy termination is considered, all the choices are bad. The problem becomes one of choosing the least unfavorable option.

For couples who choose not to terminate an affected fetus, prenatal diagnosis provides information of great value in managing their situation. In some rare conditions, such as urinary bladder obstruction, prenatal treatment is possible.

CHAPTER 8

The Crystal Ball

GENETIC ENGINEERING

The prospect of knowing the entire human genetic blueprint, or just knowing small relevant parts of it, brings up the possibility of changing human makeup. For some, this holds the cherished promise of repairing a genetic defect. For others, it looms as a sinister specter of the ultimate control of human destiny. What many have failed to realize is that gene transfer has been going on in nature all along. Sometimes it is harmful, as in the transfer of a gene from animal to virus to make an oncogene capable of causing cancer. On the other hand, the entry of a single-cellular organism into the cells of higher organisms during evolution produced the mitochondrion. It is the powerhouse for human cells.

Gene transfer has long been a scientific tool in microbiology and genetics, and it is already being used in animals. It is simply a matter of time until gene manipulation is a reality for humans. Human gene transfer has been used in the correction of severe combined immunodeficiency owing to adenosine deaminase (ADA) deficiency. Gene transfer will undoubtedly first be used to correct severe genetic defects. What

77

is to be classified as a "defect" will undoubtedly provoke debate. Reason and compassion in handling complex problems must prevail in a free society.

CLONING

Growing new plants by rooting and grafting of cuttings is an important part of agriculture. Growth of complete plants from single cells of an existing plant—cloning—is also used. Cloning has been successful in frogs by exchanging the nucleus from an egg with the nucleus from an adult animal. Embryo transfer in cattle has allowed the increased production of superior livestock. Test-tube fertilization and embryo transfer are useful in treating infertile human couples as well. The cloning of a human analogous to that of frogs could be possible down the road. However, just because it is possible to perform, should it be done? That same question is being asked about other procedures. It should be asked—and answered with reason. Our rapidly advancing technology must not be allowed to outstrip our ability to use it in a rational manner, nor should progress in research be hindered by the more deliberate pace of societal development.

Glossary

AFP (α-fetoprotein): A protein made by the fetus, but present in very small amounts in adults. Increased amounts are found in amniotic fluid of fetuses with anencephaly or spina bifida.

Amino acids: Small chemical building blocks that join together to form proteins. Three-letter codes in DNA specify the amino acids in proteins.

Amniocentesis: Needle puncture through the abdomen into the pregnant uterus to obtain fluid in which the fetus floats.

Amnion: The thin tissue of fetal origin that lines the fluid-filled cavity containing the fetus within the uterus.

Anencephaly: One of a series of defects in which the tube of tissue destined to form the brain and spinal cord fails to close, in this case resulting in absence of brain, skull cap, and scalp.

Autosome: A nonsex chromosome; one of those numbered 1–22, rather than the X or Y sex chromosomes.

Carrier: A person who has one normal and one abnormal recessive gene at a locus; a heterozygote.

Chorionic villus sampling (CVS): Obtaining fetal tissue (chorion frondosum) destined to develop into the placenta, performed at 9–11 weeks using a thin plastic tube inserted through the birth canal.

Chromatid: Each of two duplicate strands of a chromosome seen during cell division and destined to separate into each daughter cell.

Chromosome: Rod-shaped dark-staining structures visible during cell division containing genes encoded in DNA lined up like beads on a string.

79

Compound heterozygote: Having two different genes at a locus, both genes being different from each other and different from the normal, e.g., Hemoglobin SC.

Congenital: Present at birth, not necessarily inherited.

Consanguineous: Related by having a common ancestor.

Crossing over: The normal process of exchange of parts of the two chromosomes of a homologous pair during meiosis.

CVS: *See* chorionic villus sampling.

Cytogenetics: Microscopic study of chromosomes and how changes affect individuals.

DNA: Deoxyribonucleic acid, the chemical structure of genes within chromosomes.

DNA fingerprints: Fragments of DNA separated into a pattern that is unique to an individual.

Dominant: A trait requiring only one gene for expression in a person.

Electrophoresis: Separation of molecules using an electric current.

Embryo: First eight weeks after conception.

Fetus: Product of conception after eight weeks.

Gene: Smallest unit of inheritance consisting of a segment of DNA coding for a chain of amino acids (a polypeptide or protein).

Genetic code: Sequence of three bases (adenine, thymine, guanine, or cytosine) in DNA, specifying the amino acids in a protein and often written as a chain of letters, e.g., AATCGGCTT, and so forth.

Genetic counseling: Discussion of information with a person(s) about risks for, consequences of, and management of genetic disorders.

Genome: A complete set of DNA.

Germ cells: Gametes. Sperm in males, ova in females, each containing a single set of 23 chromosomes.

Hardy-Weinberg equation: $p^2 + 2pq + q^2$; where p and q are genes at a locus, then the three terms are the frequencies of homozygous normal (pp), carriers (2pq), and homozygous affected (qq). The carrier frequency can be calculated from the frequency of affecteds.

Hemoglobin: Oxygen-carrying protein of red blood cells.

Heterozygous: Having two different genes at the corresponding loci on the two chromosomes of a pair.

Histocompatibility: Inherited differences in tissues that can cause transplanted organs to be rejected.

Homologous: Chromosomes or parts of chromosomes having the same genetic loci.

Homozygous: Having two like genes at a locus.

Imprinting: A reversible change in gene action rather than a change in the DNA code itself that may be transmitted from one generation to another.

Inherited: Transmitted from parent to child in the genes.

Inversion: A part of a chromosome with reversed gene order.

Karyotype: Picture of chromosomes from a single cell arranged in order from largest to smallest.

Kindred: An extended family pedigree.

Klinefelter syndrome: A male with one Y and two or more X chromosomes, most often having 47 total chromosomes.

Linkage: Genes at loci so close on the same chromosome that they are not usually separated by the process of crossing over during germ cell formation.

Locus: The position of a specific gene on a chromosome.

Meiosis: One duplication of chromosomes followed by two cell divisions producing a germ cell with a single set of chromosomes.

Mitochondria: Small bodies in the cell cytoplasm that provide energy.

Mitosis: Process of cell division in somatic cells (cells other than germ cells) yielding two new cells each having an identical set of 46 chromosomes.

Mosaic: An individual with two or more cell types differing in gene or chromosome content, e.g., 45,X/46,XX in a person with Turner syndrome having only one X chromosome in some cells and two in other cells.

MSAFP: Maternal serum α-fetoprotein.

Multifactorial: Caused by added effects of multiple genes plus one or more environmental factors.

Mutation: Change of an existing gene to a new gene by altered DNA coding.

Nondisjunction: Failure of the two chromatids of a chromosome to separate during cell division so that both go to the same daughter cell.

Oncogene: A normal animal gene that on transfer to a virus can have increased activity in cancers when reintroduced into animals.

Ovum: A female germ cell, or egg, containing a single (haploid) set of 23 chromosomes.

PCR: Polymerase chain reaction; a method for gene duplication in the laboratory providing a sensitive, rapid method for study of genes for which the DNA code, or part of the code, is known.

Pedigree: Diagram of family relationships showing persons affected by a genetic disorder.

Polygenic: Caused by additive effects of multiple genes at different loci.

Polymorphism: Two or more different genes for a function occurring at a rate of one in a hundred or greater for the least common gene.

Polyploidy: Multiple sets of the number of chromosomes in germ cells (23) beyond the normal 46 for tissues other than germ cells, e.g., triploid = 69.

Propositus: Proband; person through whom the family is discovered.

Recessive: A trait expressed only when both genes are abnormal, either the same (homozygous) or two different genes, both abnormal (compound heterozygous).

Restriction endonuclease: An enzyme that cuts DNA only at sites for a specific short sequence of the DNA code.

RFLP: Restriction fragment length polymorphism; inherited variation in size of pieces of DNA detected after cutting with site-specific enzymes.

Sex linked: *See* X linked.

Sibship: A set of brothers and sisters.

Somatic: Cells of the body other than germ cells.

Sonography: Pictures made using echoes from very high-frequency sound.

Sperm: Spermatozoa; male germ cells containing a single (haploid) set of 23 chromosomes.

Spina bifida: Open spine, one form of major central nervous system defect (*see* anencephaly).

Syndrome: A group of features occurring together in a pattern sufficient to define a diagnosis, e.g., Down syndrome owing to Trisomy 21.

Teratogen: An agent that can cause birth defects by interfering with development during pregnancy to produce permanent change in an embryo or fetus.

Translocation: Joining of part of a chromosome to another.

Trinucleotide repeats: Long chains of three bases, e.g., CAG, CAG, CAG, . . . and so forth, in DNA.

Trisomy: Having three chromosomes instead of a pair.

Turner syndrome: A female with abnormal ovaries, short stature, and often other congenital abnormalities most often caused by having only one normal X chromosome, but sometimes other chromosome variations.

Twins: Two individuals resulting from a single pregnancy; fraternal (dizygotic) if from different fertilized eggs, identical (monozygotic) if derived from splitting of a single fertilized egg into two embryos.

X linked: Sex linked; genes on the X chromosome unpaired by the much shorter Y chromosome. One abnormal gene produces disease in males, e.g., hemophilia.

Zygote: A fertilized egg.

A Brief
Annotated Reference List

Buyse, M. L., ed.: *Birth Defects Encyclopedia*, 2 vols., 1892 pp.; Center for Birth Defects Information Services, Inc. in association with Blackwell Scientific; Cambridge, MA; Scarborough, Ontario; Carlton, Australia; Oxford, England; 1990.

A two-volume reference self-described in a subtitle as "The comprehensive, systematic, illustrated reference source for the diagnosis, delineation, etiology, biodynamics, occurrence, prevention and treatment of human anomalies of clinical relevance." It is multiauthored and authoritative. Entries are listed alphabetically, as are alternative names, often requiring switching from volume to volume in a search. The Birth Defects Information System for computers keeps this information up-to-date and makes searching easier.

Gelehrter, T. D. and Collins, F. S.: *Principles of Medical Genetics*, 324 pp., Williams and Wilkins; Baltimore, Hong Kong, London, Sydney; 1990.

A general textbook of medical genetics containing many excellent illustrations with captions that clearly explain complex principles.

Jones, K. L.: *Smith's Recognizable Patterns of Human Malformation*, 778 pp.; W. B. Saunders; Philadelphia, London, Montreal, and other cities; 1988.

Genetic disorders producing malformations are succinctly described with illustrations. It also includes sections on principles of genetics and mechanisms of malformations. Helpful lists of malformations and associated syndromes supplement the index.

McConkey, E. H.: *Human Genetics: The Molecular Revolution*, 322 pp., Jones and Bartlett; Boston, London; 1993.

A readable text of human genetics with excellent summaries and references at the end of each chapter.

McKusick, V. A.: *Mendelian Inheritance in Man*, 11th ed., 2 vols., 3009 pp., The Johns Hopkins University Press, Baltimore and London,1994.

This is an encyclopedia of human genes and genetic disorders. It describes genetic disorders caused by single genes, and includes historical and current references. It contains gene maps and catalogs of autosomal dominant, autosomal recessive, X chromosome, Y chromosome, and mitochondrial genes. The constantly updated on-line version (OMIM) is available by computer as part of the Human Genome Project through the Internet service:

In the United States   help@gdb.org
In the United Kingdom   admin@hgmp.mrc.ac.uk
In Australia   reisner@angis.su.oz.au

OMIM is also available via the World Wide Web (WWW) through the GDB home page (http://gdbwww.gdb.org).

A 1995 CD-ROM version (PC-GDB and OMIM) is also available. Both OMIM and the CD-ROM have added brief clinical synopses and provide computer search facilities. It is a primary reference for additional information on specific disorders.

Milunsky, A.: *Choices, Not Chances*, 488 pp.; Little, Brown; Boston, Toronto, London; 1977.

A guide for questions about heredity and health. It sensitively presents many illustrative case histories.

Schinzel, A.: *Catalogue of Unbalanced Chromosome Aberrations in Man*, 913 pp., Walter de Gruyter; Berlin, New York; 1984.

This is a listing of human chromosomal abnormalities by chromosome number with illustrated clinical descriptions of the resulting disorders.

Scriver, C. R., Beaudet, A. L., Sly, W. S., Valle, D., eds.: *The Metabolic and Molecular Bases of Inherited Disease*, 7th ed., vol. 1, 1652 pp.; vol. 2, 3177 pp.; vol. 3, 4605 pp.; McGraw-Hill; New York, London, Toronto, and other cities world-wide; 1995.

This is a huge multiauthored compendium of metabolic disorders, which goes far beyond what the title suggests, including sections on cancer genetics and basic principles. If there are questions about metabolic diseases, look here.

Thompson, M. W., McInnes, R. R., and Willard, H. F.: *Thompson and Thompson: Genetics in Medicine*, 5th ed., 500 pp., W. B. Saunders, Harcourt Brace Jovanovich; Philadelphia, London, Toronto, Montreal, Sydney, Tokyo; 1991.
    A general textbook of medical genetics containing explanations of genetic principles, illustrated descriptions of genetic disorders, a glossary of terms, and references.

Index